What Is Your Cat Saying?

What Is Your Cat Saying?

Dr. Michael W. Fox & Wende Devlin Gates

Coward, McCann & Geoghegan, Inc.
New York

For Sam and Mocha
M.W.F.

For my parents, Wende
and Harry Devlin
W.D.G.

Photo Credits

Joan Ehrlich: Title page, 55-57, 60, 63

Constance Fogler: 24

Sumner Fowler: 32

Michael W. Fox: 6, 8, 12, 17, 19, 26-27, 30, 34, 38, 40, 42-44, 46-47, 49, 53, 58 (lower), 65, 74-75

Humane Society of the United States: 14 (Cima), 25 (Dantzler), 33 (Spies), 39 (McNees), 51, 58 (upper/Lepine), 62 (Baker), 64 (Dommers), 68

Bonnie Smith: 77

Nanette Stevenson: 28

Copyright © 1982 by Michael W. Fox and Wende Devlin Gates
All rights reserved. This book, or parts thereof,
may not be reproduced in any form without permission
in writing from the publishers. Published simultaneously
in Canada by General Publishing Co., Limited, Toronto.
Library of Congress Cataloging in Publication Data
Fox, Michael W.,
   What is your cat saying?
   Includes index.
   Summary: Questions and answers explore various aspects of cat behavior and examine methods of choosing and training a kitten.
   1. Cats—Juvenile literature. 2. Cats—Behavior—Juvenile literature. [1. Cats. 2. Questions and answers] I. Gates, Wende II. Title.
SF445.7.F69   636.8   81-4884
ISBN 0-698-20443-3   AACR2
Designed by Marion Mundy
First printing
Printed in the United States of America

Contents

1. What Is Cat Talk?
   Questions

2. How Your Cat Got to Be That Way
   Questions

3. Equipment for Cat Body Language
   Questions

4. Cat Body Language, or "Felinese"
   Questions

5. Bringing Up Your Kitten
   Questions

6. Fun and Problems
   Questions

   Index

1.
What Is Cat Talk?

What is the big cat saying to the kitten in the picture on the left? The big cat's body movements and hissing sounds make the message very clear to the little one: "Watch out! Move away from me and don't come near again or I may scratch and bite you!"

The big cat is unfamiliar with the kitten's smell, so it shows with its body and voice that the kitten is not welcome. The kitten responds by arching its back defensively and moving its body away from the threatening older cat.

You can probably tell immediately what is happening between the two cats just by looking at the picture. You are able to "read" their body language and figure out what they are saying to each other because the postures of these cats are easily understood.

There are, however, other body movements and patterns of behavior which are not always so easy to figure out. This book will, we hope, give you an understanding of these systems of communication so that you can soon tell what your cat is saying in almost every situation. You'll find your relationship with your cat will become more exciting once you learn the cat's system of "talking." Communication between you and your cat can become rich indeed!

Learning cat body language is more difficult than learning the language of dogs. Even though you may already own a cat, and may be familiar with some of its movements and what they mean, much of cat behavior and body movements is not easily understandable. There are two main reasons for this.

First, there has been very little serious research done on cat behavior until recently. Dr. Fox has spent much time observing and researching cats. He has been able to provide us with a great deal of information on cat behavior.

Second, cats are less dependent on human actions than are dogs. A dog is more responsive to a leader or master. This makes a dog more trainable. Most members of the cat family are not as gregarious as dogs and are able to exist on their own in the wild. Their behavior is therefore less predictable or obvious compared to an exuberant dog. Since it is often hard to tell what cats will do next, humans have frequently seen cats as mysterious creatures, and many strange superstitions have grown up around them.

Cats are less domesticated than dogs. They seem to have the best of both worlds. They manage to enjoy the benefits of a warm, cozy house and daily meals while still retaining their ability to run out into the woods, to hunt and kill prey, and to survive without humans if necessary.

Because cats are so different from dogs and other domesticated animals, watching them is especially fascinating. Once you learn about cat talk, you'll have an even greater appreciation for these exquisite and very special animals.

Questions

Q.   Don't owners have any effect on their cats? If we don't pet and talk to the cat, or if we really mistreat him, will he turn out mean? A dog would become mean if he was treated badly.

A.   Our behavior and body language do have an effect on cats. Like babies and pups, cats need handling and touching. They like to be talked to and played with. Unlike a dog, however, cats will not react as strongly to your actions. A scolded cat may have a fleeting submissive reaction or short period when it seems fearful of you. A dog, however, may sulk for hours. If badly mistreated, a cat may eventually become fearful and will scratch or bite when animals or humans come near.

Q.   Don't cats have to learn their body language the way human babies do—from their parents?

A.   Cats do learn some of their language from their mothers but not nearly as much as human babies do. Cats are born with the ability to move and behave the way they do without having to be taught by their parents. Or, in scientific terms, most of their behavior is "innate."

If a cat lost its parents, it could survive alone in the world more easily and at a younger age than a dog could. Most cats know how to hunt and kill their prey and can climb trees and fend for themselves from an early age. They are "programmed" to be independent.

One reason that dogs extend their childhood and puppy-like behavior is that they are "pack" animals. In the wild, wolves and dogs run in packs and help each other hunt food. Being dependent rather than independent helps them stay in a pack. Wolf cubs and puppies can enjoy a long childhood.

Cats are relatively solitary animals. In the wild, they are often rejected by their mothers by the time they are six to eight months old and have to fend for themselves, hunting alone for food and forced to act like adults early in life. Only one member of the cat family, the lion, lives in a social group —the pride.

2.
How Your Cat Got to Be That Way

The cat has had a colorful history. Its ancestors have been treated as evil animals sent from the devil, or revered as gods, their coffins decorated with emeralds, gold and carved ivory. Scientists believe that the cat is descended from the Kaffir cat, a wild cat easily domesticated and easily bred with domestic cats. And of course your proud feline is distantly related to such magnificent beasts as the leopard, the stealthy panther, the fascinating lion family and the fastest animal on earth, the cheetah. We also know that cats have qualities that seem to separate them from all other animals in the world. Their superior physical abilities—speed, grace and balance—make humans fear and respect them. Cats also possess unusual beauty.

If you visit a zoo to look at your cat's relatives, you can study their body language. Unfortunately, they can't be free to behave the way they do in nature, but if you watch the large animals in motion, you are seeing a bigger version of your own cat's movements.

And watching your cat in your own home is almost the same as watching it in the wild. This is because cats still behave naturally or instinctively, even though they are domesticated.

In this chapter we'll examine the background of your pet and see how your cat evolved.

Questions

Q.   *Were cats really buried in fancy coffins?*

A.   Yes! Cat mummies buried in beautifully decorated coffins have been discovered in Egypt. Cats appeared in sculptures, tomb frescoes and paintings as early as 2500 B.C. Remains of domesticated dogs have been found dating from 14,000 B.C.

It is believed that cats were originally brought from Libya and Nubia to Egypt, possibly to keep the rats out of granaries. These cats were also trained to hunt birds and fish. They soon became symbols of grace, fertility and femininity. Egyptians began to worship a cat-headed goddess called Bastet. When a house cat died, the owners were required to shave off their eyebrows as a symbol of mourning.

Apparently cats were also found in China and Japan. They were used to protect the barns from mice and rats. The domestic cat arrived late in Europe. Cats were so honored that citizens were forbidden to export them. It is thought that they were finally smuggled into Europe on trading boats from Egypt.

Q.   *When did cats come to America?*

A.   Not soon enough! When rats started parading down main streets, our ancestors finally decided to import some help. Cats probably came from England to America in the seventeenth century, to cities such as Philadelphia. The rat problem was soon brought under control.

Q.   I heard from two friends that if a black cat walks in front of you, you'll have bad luck. Is this true?

A.   No. So many silly superstitions about cats still exist. Several hundred years ago, there was a superstition that if a cat sneezed, it would rain. There were some cruel superstitions too. People believed that to seal a cat up in a wall would bring a new building good luck, or that if a farmer buried a live cat a good harvest would result.

In the Middle Ages, a group of people who were said to practice witchcraft associated themselves with the mysteriousness and fertility of the cat. This pagan cult was found in Northern Europe and was called Freyia. They and their cats were hunted and often killed. During the Renaissance, the Catholic Church under Pope Innocent VIII ordered the destruction of cats. Since witch-hunting meant cat-hunting, cats were also persecuted in the New World—in South Carolina and New England.

The cat had become a symbol of evil. This was true even after cats brought the black rat under control. The black rat was the vermin that carried the terrible Black Death, a plague which wiped out half the population of fourteenth-century Europe.

Superstitions about cats are passed on from generation to generation. Some people still see cats as evil, distant and unpredictable. Cats often respond to people's fearfulness about them by being jumpy and even aggressive. This perpetuates the fears, and so the cycle continues. We hope this book will help everyone understand cats and will abolish the superstitions surrounding them.

Q.   How did cats survive those witch-hunts?

A.   Cats survived because people started to realize that they are the only natural enemy of the rat. When the brown rat swept through Europe in the eighteenth century, carrying disease, eating food in homes, stores and warehouses, and frightening whole villages, there was a sudden new respect for the cat. Good ratting cats fetched a fine price, and they finally brought the rat population under control.

In the nineteenth century, Louis Pasteur, the scientist, discovered that tiny microbes were the carriers of many diseases and were associated with dirt. People started to fear that all animals might carry disease because they were dirty—all animals, that is, except cats. Cats licked themselves clean and used one chosen spot for their toilet. People came to feel that the cat was the one animal they would be able to keep in the house without catching all kinds of diseases. Once again, the cat survived!

Q.   I love our white tomcat Billy. I don't understand how cats became part of superstitions and witch-hunts, especially since they helped so much all throughout history. Why were cats singled out?

A.   Good question! People often have strange attitudes toward animals, and for that matter toward anyone who has special abilities. If cats have special talents that are helpful, some people believe they may have talents that are harmful as well. Since cats, unlike dogs, can't really be controlled, it is sometimes thought that they might do something unpredictable and harmful. People are fearful of the unknown, and to some the cat seems a mysterious and wild animal that cannot be controlled.

One superstition warns mothers about ever letting a cat near a baby because the cat might choke the baby by trying to lick the milk in its mouth. If you study the body actions, language and behavior of a cat, you will know that its range of action is limited to hunting only natural prey, such as mice, rabbits and moles. Only an emotionally disturbed or badly mistreated domestic cat would viciously attack a human.

Q. *At the zoo today, we saw a family of lions. They were so huge! And they're from the same family as our kitten. Why are the members of the cat family so different in size?*

A. Cats have evolved to different sizes to adapt to the size of the prey they hunt. Cats of varying sizes can then live in the same area so they won't compete for food. The smaller cats, such as the bobcat, may hunt only rabbits, mice and lizards, while the mountain lion will leave those prey alone and concentrate on deer and boar. Each cat has its place in the ecology of a given area.

The lions you saw are the only real social group in the cat family to hunt in a pack. The maned lion is the head of the pride. Although he has a ferocious reputation, he's a kitten compared to the lionesses who do all the hunting for this lazy male. The maned lion spends most of the day sleeping and roaring and marking the pride's territory. When his "queens" or lionesses make a kill, *he* eats first, the lionesses next, and the cubs last. If there is not enough food to go around, the cubs are the first to starve and die. Although it may seem cruel to allow cubs and weaker adults less food, it controls the population of the pride. Lions breed so quickly that if there were no population control, all their food sources would soon be wiped out.

Q. *Why don't cats help each other and hunt in packs like the lion or the dog?*

A. Cats don't need each other to hunt down prey. They have all the abilities they need to survive by themselves. Also, since they hunt small prey such as mice and insects, there wouldn't be enough to share if more than one cat were to hunt together. A solitary cat can calculate how close to get in order to rush, leap at and pull down its prey. Once it catches its prey, it kills with one bite which severs the spinal cord of the animal. The cat's ability to ambush its prey by using stealth serves it well, since it is not built to run after its prey like the wolf. The cheetah is the one exception here in the cat family. Cheetahs can run like the wind for relatively long distances.

Q. *Isn't there any way to track down the wild cats so we could learn more about their behavior?*

A. Yes. People who study animal behavior are now using a handy tool called the biotelometer. This tiny transmitter is attached to a collar and placed around the neck of a captured animal. It is then easy for scientists to follow the animal and watch how it lives. Wild cats are very important to the ecology of each area they live in. We should try to stop people from trapping wild cats like the lynx and bobcat for their beautiful fur. Also, pet cats can get caught in these traps and may be killed, so that their pelts too can be used to trim hats and coats. Not only will such trapping end the life of beautiful wild creatures but it will destroy the ecosystems of the areas in which they live.

Q.   We saw a stuffed saber-toothed tiger in the Museum of Natural History in New York City. It was frightening. It was enormous and had nine-inch teeth! These tigers are now extinct. What happened to this part of the cat family? Are there any big cats like the saber-toothed tiger which have survived?

A.   One cat which looks very much like the saber-toothed tiger remains. This is the clouded leopard in the photo below. The ferocious saber-toothed tiger, which lived in prehistoric times, ate mainly mastodons, mammoths and elephants. One explanation for its extinction is that once humans became the hunters of these huge, slow-moving animals, the saber-toothed tiger died out. The oldest ancestors of cats are thought to be civets and mongooses. Cat fossil remains have been traced back thirty-five million years.

Q. My cat has "tufts" on each ear, just like a lynx. Do you think he's really part lynx?

A. Your cat probably has no lynx blood in him, since domestic cats are not directly related to the lynx. Many domestic cats have these pretty tufts. They make the ears look larger so that ear signals can be more clearly seen. Ear position is very important in understanding what your cat is saying.

Q. I've heard two interesting things about cats: White cats are deaf; and there is no such thing as a male calico cat. Are these true?

A. White cats and white animals, which are known as albinos, are often deaf, but not all white cats cannot hear. Often a young cat or kitten doesn't respond to your voice and so may seem deaf, but it may become more attentive as it grows older.

Male calico cats are rare and they are usually sterile. Most calico cats are female.

Q. Aren't Siamese cats supposed to be cross-eyed and have a kink in their tails in order to be considered real thoroughbreds?

A. No. Many Siamese cats are indeed cross-eyed, have squints and have crooked vertebrae in the tail, but these traits should be avoided when buying a Siamese. They are often the result of overbreeding (lack of quality control) and inbreeding (breeding the cats among members of their own family— mother with son, brother with sister, and so on). Thoroughbred Siamese cats should not have these physical problems.

3.
Equipment for Cat Body Language

Cats are extraordinary animals. Their bodies are marvelously equipped for survival. As solitary hunters, they cannot rely on others to act as guards or scouts to offer a helping paw, as dogs do. Everything must be done alone.

The cat is equipped with *nine* senses—smell, touch, taste, temperature, balance, sight, hearing, time and direction. All these senses work in conjunction with a highly evolved brain. On the evolutionary scale, the cat's brain is halfway between the most highly developed brain of the ape and, at the lower end of the scale, the brains of rats and mice. It has a well-developed emotional center and is able to experience or sense emotions as we do.

When you observe cat body language, you see many of the cat's senses being used. While watching your cat in the yard, you'll see it suddenly stare, crouch and prepare to leap. *You* may have seen and heard nothing, but your cat's eyes have an unusual structure which allows them to see any movement over a large area. Its sense of sight is remarkable.

Its sense of direction is also remarkable. One newspaper told the story of a lost cat traveling two hundred miles to its home. There have been other stories about cats finding their way home from as far away as one thousand miles. Or, even more amazing, cats who find their owners in places the cats have never been before! The cat's body structure allows it to do these remarkable things. Understanding its senses and abilities will help you understand what your cat is saying to you, to other cats and even to itself.

Questions

Q. *I think the most amazing thing about cats is their ability to land safely on their feet when you drop them. Why can't a dog or other animals do this?*

A. This "righting reflex" is part of the cat's legendary sense of balance. Incidentally, not all cats will land safely on their feet. Test your cat by holding its forelegs and hind legs three feet above a pillow and then letting go. A large slow cat may not respond to this test at first. And of course don't ever drop your cat from a height higher than five feet. Although some cats can survive a twenty-foot fall from a tree, not all will. How can the survivors do this? As a reflex action, the cat twists until it's right side up and extends its resilient legs, which act as a landing pad. These movements reduce the chance of injury or internal damage.

Q. *How is the cat able to walk and balance itself on a fence? A dog could never "tightrope-walk."*

A. Actually, the wolf and some domestic dogs can walk a slender ridge. But most dogs don't have the finely developed sense of balance a cat has. As tree climbers, cats have a highly developed part of the brain known as the cerebellum, as well as a balancing system in the inner ear. This is where our sense of balance is also located. The cat's tail may also help it balance itself.

Q.   You talked about a cat's ability to see movement in a yard. I know my Abyssinian cat can spot a chipmunk thirty feet away. How does this work?

A.   Because of the structure of its retina, the cat doesn't take in every detail in its sight. Instead, the eyes are tuned to detect movement. This is why a mouse or other animal will sometimes "freeze" so their predators don't see their movement. The retina at the back of a cat's eye has more rods than cones. Rods are cells that are sensitive to light; cones are sensitive to color. Cats don't need to see color as much as they need to see in the dark for night hunting.

Q.   My kitten Trixie's eyes look funny when I pet her. Why is this?

A.   When your cat is relaxed, a fold of skin called the membrana nictitans, which is normally folded into the inner corner of the eye, moves partway over the eye. This membrane also protects a cat's eyes from dust and dirt when it chases after prey in bushes and fields.

If you ever discover Trixie's membrana partly covering the eye *without* your petting her, this may be a sign that she is sick.

Q.   Why do a cat's eyes glow when a car's lights shine on them in the dark?

A.   Cats, as well as deer, rabbits, foxes and dogs, have a layer at the back of their eyes which reflects light. It is called a tapetum. All these animals are night hunters and can see in very little light. The tapetum allows more light to pass into the retina.

Q. My cat's eyes are really strange. The centers, or pupils, sometimes become oval-shaped. My pupils, and my dog's pupils, are round. Why do Little Yellow's pupils change?

A. In humans and dogs, pupils are always round. But cats have oval or elliptically shaped eyes during the day. They have very changeable pupils. You can often tell what your cat is saying by watching its pupils. When it's afraid, its pupils dilate. When it is angry, they become slits. They also become slits in bright sunlight. At night the pupils become very dilated.

Watch Little Yellow's eyes when he's concentrating on something. The pupils change shape. The pupil reflex allows a cat to judge distances, and since its eyes are on the front of its head (not on the side, like a horse's), it can tell how far it has to leap in order to catch its prey.

Q.   My calico cat's eyes close when she is being petted. In fact, she's constantly rubbing up against me asking to be petted. Why do cats want to be petted so much?

A.   First, and most of all, it feels wonderful. Dogs, people, other mammals, all *need* to be petted and cuddled. Babies and young animals can even die if they are not handled enough.

Cats do seem to have an extra-special need to be petted. From birth, cats ask to be stroked and as adults they stretch out and purr like baby kittens. There are nerves around the hair follicles which send positive messages to the cat's brain when their bodies are stroked. Therefore, petting has a calming and beneficial effect on the nervous system.

Petting also reaffirms social relationships—with other cats and with humans. When a cat asks to be petted, it's saying, "I want some pleasure and I'd like to cement our relationship." Some cat owners get upset when they see that the cat asks for petting from just anyone who comes in the door, not simply the person it knows best. This is a cat's way of greeting and accepting a new person.

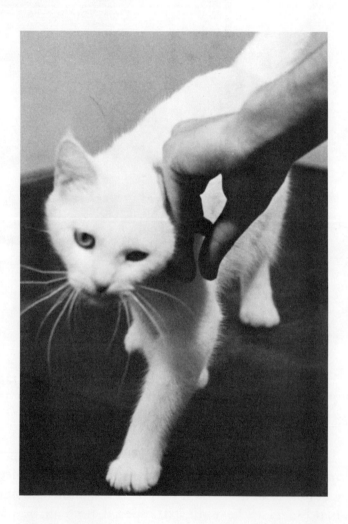

Q.   I saw a stray cat smelling the droppings of another cat. The stray opened his mouth and seemed to breathe in the smell. What was he doing?

A. Cats have a strong sense of smell. The sense of smell is often assisted by an organ called the "vomeronasal" organ, which is connected to the mouth by two ducts in the hard palate. Cats open their mouths to enable the smell of certain substances, such as catnip, to reach the vomeronasal organ. This is called the "Flehmen response."

Q. Do cats have better senses than dogs? How about hearing? Don't dogs have better hearing ability than cats?

A. At high frequencies both the dog's and the cat's ability to hear are much better than a human's. Some say cats can hear higher frequencies than dogs. They can also discriminate one-fifth to one-tenth of a tone nearby. This is very important for hunting. Cats have a wonderful advantage over many other animals by having a moving external ear called the pinna which can be used to collect sound waves all about the head. Cats also can tell the direction and the distance of sounds by feeling how loud and how strong the noise is when it is received by each ear.

Q.   *I know what my big cat Mr. Puss is saying when he runs in from two rooms away and jumps on the food counter as I start opening* any *can. Does the sound of a can being opened have any particular effect on a cat's system, like the sound of a high-frequency whistle for a dog?*

A.   When Mr. Puss rushes for the kitchen at the sound of an opening can, it naturally means "I'm hungry, I want dinner." He associates the sound of the can being opened with filling his stomach. This is something he has learned. A knock on a door will also excite a cat because this signals a new arrival. This too is learned. Mr. Puss has probably learned his name as well. However, certain sounds don't have to be learned. Something like the rustling of cellophane will excite a cat because it sounds like a mouse or other small prey that a cat might chase instinctively if it heard rustling in the grass.

Q. *My cat seems to know what I'm planning to do. When I'm getting ready to visit my friend, she jumps on my suitcase and won't move. How does she know what's going on?*

A. Cats are great observers of human behavior. They seem to know all your routines. They know when you're unhappy, when you're going out to play or leaving for a weekend. Some people call this talent ESP or extrasensory perception.

Q. *My cat goes to the windowsill exactly two minutes before my father comes home from work, though he arrives at a different time every day. How does she know?*

A. The cat hears and recognizes the footsteps or sound of the car outside way before you do. One owner was awakened by a cat who ran up and down his bed. When the owner got up, the cat charged at the door. They went outside only to have the house collapse in an earthquake behind them. The cat probably sensed slight tremors in the earth minutes before the earthquake finally arrived. Other animals can apparently sense these tremors too. Scientists are now studying the behavior of certain animals to help predict coming earthquakes.

What is amazing is a cat's sense of time. One of Dr. Fox's cats knows his master gets up at precisely 8:00 A.M. every morning and licks him two minutes before. This is hard on Dr. Fox on weekends, when he would sometimes like to get a little extra sleep. Fortunately, his cat finally adjusted to daylight saving time!

Q.   Why is the sense of time so important in cats?

A.   The sense of time is important to cats who live in the wild because they have to avoid other wild cats who compete with them for food in the area. Cats can usually avoid each other's territory by staying away from freshly sprayed trees and bushes, since spraying is the way cats mark their territory. By smelling the sprayed scent, another cat knows to stay away until a later time. If a spray is recent, the cat can adjust its internal clock and will return when its rival has probably gone. A sense of time also enables a cat to know when to go out hunting, and it may also help in keeping a sense of direction so that the cat does not get lost.

The lion is spraying the tree to mark his territory, just as domestic cats do.

Q.   How can cats who get lost while they are away from home find their way back again? Isn't this ability a little exaggerated?

A.   No. As we said, cats have a very sharp sense of time. They have a physiological clock inside them which is set to their local time. If they are away from their home, this internal clock will not correspond to the sun's position in a place, say, one hundred miles from home. Cats can determine the time difference in order to figure out the direction toward home. The cat's senses work somewhat like those of a homing pigeon. The ability to tell time and direction are very much connected in these animals.

Q.   What about the stories of lost cats finding their owners in a new home in a neighborhood or state where they've never been before? That's really hard to believe.

A.   It *is* hard to believe. One owner who happened to be a veterinarian left his cat in New York when he and his family moved to California. Imagine his surprise when, six months later, he saw his cat, thin and exhausted, crawl into his house one morning. He didn't believe it at first. He thought this animal simply looked like his original gray cat. But he was able to establish positive identification by feeling the unusual bump on the bone of its tail.

This remarkable ability is called "psi-trailing." And we're not sure how cats or other animals are able to locate their owners. More research is needed on this fascinating subject.

Q. *I'm afraid of cats. When my friends and I visit John Phillip, who has two cats, these pet cats seem to rush right up to me. They actually single me out. I really get scared. Why do they pick on me?*

A. It is so common for a cat to rush up to someone who is afraid of cats that there must be a reason for it. Your fear and anxiety may be creating a negative energy field, which attracts cats, or your passive "freezing" may be interpreted as a friendly signal. We don't yet have a definite answer to this question.

Q. *When I take a bath, my dog and cat stand near the tub and look as though they want to come in for a swim. The dog tries to lick the water and jump in. Even the cat seems to want to join me. I thought cats didn't like to swim. Do they?*

A. Cats can swim. Tigers, for instance, love to swim in pools, and bobcats are very good swimmers. Domestic cats are good swimmers too, but they don't like to get wet because they don't have oily coats as dogs do, and therefore feel the water more. The fact that dogs also have a film of guard hairs that keep their undercoat dry may explain further why your dog seems more eager than your cat to jump into the tub.

Q.   My friend Christopher's cat closes its eyes when I touch its whiskers. What are whiskers for, anyway?

A.   Dr. Fox believes that whiskers, which are extremely sensitive to touch, help a cat tell from which direction a smell or breeze is coming. Watch the body language of Christopher's cat next time it seems to be hunting for something. It will twitch its whiskers as it sniffs around.

Whiskers also seem to protect a cat's eyes. This is why Christopher's cat closed its eyes when you flicked its whiskers. When a cat travels through bushes, the whiskers probably protect its eyes from being scratched.

Q.   My dog's tongue is smooth. Todd's kitten has a very rough tongue, almost like sandpaper. Why is the kitten's tongue so rough?

A.   The better to lick clean the bones of mice, birds and other prey! Todd's kitten also needs its raspy tongue to groom itself. Incidentally, sometimes the loose hairs on a cat are swallowed. Cats usually pass these fur balls, but if they can't, they may be given a little mineral oil, a mild laxative or even grass to induce vomiting.

4.
Cat Body Language, or "Felinese"

In this chapter we'll discuss the actual body movements of cats and how they are associated with different behaviors. These are: marking behavior; fearful and aggressive behavior; social investigation behavior; vocalizations; sexual behavior; maternal behavior; hunting behavior; and play behavior.

As we said earlier, cat behavior differs from dog behavior in two important ways: cats are more solitary and not group- or pack-oriented like dogs; cats may not have been as affected as dogs by domestication. Cats display relatively less interaction with each other and very little group behavior. In fact, much time in the wild is spent in *avoiding* each other. As a result, social interactions of any length are normally limited to chance encounters, sex or rivalry fights between two males. Also, two cats do not both act dominant and aggressive at the same time, as dogs do. While dogs tend to have a fixed dominance hierarchy, cats have what is called a "relative" dominance hierarchy, which can depend on time and place. For example, one cat may be dominant over another at feeding time, while the other one is the boss when it's time to choose a sleeping place. Another fascinating aspect of social behavior among cats is the occasional "cat party." These gatherings usually happen in the same spot at irregular intervals. More about that later.

Knowing what your cat is saying means being able to pick out its body movements in many different situations. Understanding behavior patterns will also help you sort out what your pet is saying. With a little careful observation and a little study, you'll be speaking "felinese" in no time!

Questions

Q.   I'm not sure I understand "relative" dominance.

A.   As we said, there is no fixed hierarchy or order of top cat, second cat and "runt" as there usually is among dogs. Cats in a neighborhood may have an order of dominance, but it is not rigid. For example, if an inferior cat has entered a common passageway, a dominant cat may let it pass. Cats regulate their traffic on these pathways by visual and olfactory (scent-mark) contact. Also, as we said, the dominance hierarchy may change in different contexts and places. When one cat so threatens another that it is left to eat first, but moves away when the other wants to sit or sleep where it is lying, then we speak of relative dominance. If such a cat had its way all the time, that would be an example of absolute dominance.

Q.   When we moved into a new neighborhood in Texas, our tomcat came home all scratched up from a fight with other cats. He'd never been in a fight before. Why did this happen?

A.   A new male cat in the neighborhood usually has to fight its way into the "brotherhood." We humans may find ourselves faced with similar situations. In school, you may find yourself trying hard to become a member of a certain group or club. The hierarchies in these cat brotherhoods are fixed (as they are in many of our school and adult clubs), and a new cat must try to find its place without disturbing the prevailing order. In the case of your tomcat, it might be kinder for you to castrate him. Castration, or surgical removal of the cat's testicles, will eliminate or at least cut down on his urge to roam and fight.

Q.   Do cats have a feeling for their home territory as dogs have?

A.   Yes. And wild, free-roaming cats have these feelings of territory also. Free-roaming cats have a territory outdoors much as your cat has its house or apartment as its territory. I'm sure you've noticed that, in addition to its home territory, your pet has a special resting spot and a special sunbathing spot as well. Cats become more aggressive about defending their home during the mating season. Aggression, territoriality and sexuality are behaviors which are all somehow related in cats.

In the wild it is necessary to defend territory in order to keep "poachers" away from the "resident" cat's prey. In addition to home ranges, free-roaming house cats have network paths, some of which lead to central areas which are used for hunting, courting, mating or partying.

Q.   Do cats really have parties? Do they really celebrate the way we do?

A.   Well, yes, cats do celebrate in a way, and they certainly communicate with each other. Imagine a bevy of males and females all caterwauling, licking and grooming each other, meowing and even hissing and flattening their ears in anger. They're all speaking in cat body language in this peculiar cat situation, the neighborhood party. These gatherings are purely social, but we don't know why they occur. Perhaps the area in which the party is held is a neutral zone. Or perhaps these parties are like our "mixers" or dances.

At other times and in other places, these cats might fight or chase each other, but here they are all peaceful and sociable. This shows that cats aren't innately antisocial or solitary. They do have get-togethers sometimes. But then suddenly *all* the cats will leave the party at the same time and go home!

Each cat in the household has chosen its own sunbathing spot.

Q.   My parents breed cocker spaniels. I love to watch the puppies. Now our Siamese female has had kittens. It's the first litter of kittens I've ever seen. They're very different from the dogs. With our dogs, the most aggressive dog at the feeding bowl was also the most dominant dog when they played. He was the "top dog" in the litter. But the most dominant kitten at the feeding bowl, a black with white stockings, is not the most aggressive kitten in play. In fact, she gets beat up once in a while. Why is this?

A.   Among kittens and cats, there is often a clear and stable dominance order at the feeding bowl. The dominance order is also rigid in other places, such as the brotherhood mentioned before, and sometimes in resting places. But there is not as clear-cut a relation between dominance and aggression as there is among dogs. So, although the dominance order persists to some degree away from the bowl, there is no aggression if the order is violated. For example, if an inferior cat did not leave a favored resting spot when a dominant cat approached it, the inferior one might be ignored rather than challenged. But among dogs, a dominant animal would more likely threaten and chase its inferior for insubordination. Pack animals must always respect their leader.

The older cat is dominant here, and the kittens' posture shows this.

Q.   What is spraying? My kitten hasn't done that yet.

A.   Spraying is marking behavior. The cat backs up, positions, holds its tail up straight, trembles and then sprays. Your kitten hasn't done it yet because cats of both sexes don't usually spray until sexual maturity. The liquid contains a fatty substance which leaves an unpleasant odor, as you may discover. We recommend that you eliminate this and a lot of other problems, such as unwanted litters and aggressive behavior, by having male kittens castrated and females spayed. In the wild, spraying behavior has many purposes.

Q.   What kind of purposes?

A.   Spraying has the same function as the urine or droppings of a dog. It serves as a "calling card," saying "I was here." It also enables cats to avoid unwanted encounters. A cat passes, sniffs and knows how recently another cat has passed by—perhaps even who the other cat is. Cats may also use spraying to mark the boundaries of their home range territory. Spraying in males increases when stimulated by the presence of a female in heat.

Cats mark in other ways besides spraying. When a cat claws your furniture, it is marking its territory—at your family's expense! Besides spraying and clawing, cats use rubbing as another way of marking.

Q.   How is the cat's rubbing against something like leaving a calling card? I don't understand.

A.   The cat has scent glands along the tail, on each side of the forehead, on its lips and on the chin region. By brushing against a chair with its tail, chin or lips, the cat will leave its scent. When a cat rubs against another cat that it knows, or against you, in a friendly manner, you may think it wants to be scratched, but in many cases it only wants to leave its scent on you.

Q.   Besides marking behavior, how else can you tell what your cat is saying?

A.   There are three basic categories of feline behavior, and all other cat body language consists of these in varying degrees of intensity:
  1. *Offensive posture*—This is when a cat is staring directly at its enemy and is poised to jump and attack.

  2. *Defensive posture*—This is when a cat shows its "Halloween" position, with its back arched, turned sideways. Sometimes the front legs are more defensively set than the back legs, and the cat will "crabwalk." The back is actually less afraid than the front!

3. *Passive posture*—This shows the cat's aggressor that it is ready to surrender. The cat crouches submissively, like a dog, although it won't roll over as a dog will in that situation.

Q. *Speaking of rolling over, I notice my dog rolls over when he's submissive. My cat rolls over too, but not because she's submissive. Why does she do this?*

A. Rolling over in cats is usually body language for "I want to mate." It can also mean "I want to be scratched" or "I want to play."

Q. *Why does my cat Edith stick her tail straight up in the air when she comes to greet me?*

A. The tail high in the sky is greeting behavior. This is active submission. A dog, in this case, would wag its tail and jump up to greet you. This "tail up" communication is derived from Edith's infancy when all the kittens in her litter would lift up their tails for their mother to lick their anal region clean and to help them evacuate. Edith is acting like a friendly kitten when she approaches you tail up.

Q.   I have ten cats. Yes, ten. And I love them all. One funny thing I've noticed is that one male cat, Henry, will mount, tread with his paws and bite the scruff of one meek male cat called Pumpkin. Pumpkin crouches in that defensive posture. Why does Henry do this to Pumpkin?

A.   This is another example of behavior derived from infancy. Henry is not trying to mate with Pumpkin but instead is showing his dominance over him. Henry is a high-ranking male and Pumpkin ranks low in the group of cats.

Q.   Who do cats sniff each other when they meet? What are they saying?

A.   When cats sniff each other on meeting, it is as if we said to another person we just met, "Hi. What's your name? Where are you from? Where do you go to school?" and so on. This is called social investigation. Cats sniff each other's scent glands at the mouth and along the anal area. They often touch each other's noses—and when they are friendly, they even rub heads.

Q.   Why does a cat turn sideways and arch its back when it's frightened or threatened?

A.   This is a body movement which makes the cat look bigger and more ferocious than it really is, like a dog whose hackles are raised and who walks stiffly on tiptoes. The male lion has a mane to intimidate its enemies. And the lynx has ear tufts. A confident cat will threaten only with its head, and a more defensive cat will turn sideways and arch.

Q. *I know one should never stare directly at a strange dog because it means a challenge. Is this true with cats?*

A. Yes. A stare from a cat is also used to keep you or another cat at a distance. But most facial expressions on a cat are not as obvious as on a dog. Cats do have certain expressions, such as an offensive or defensive threat, the Flehmen response, the passive-submissive looking away, and the face we talked about in Chapter 2 when the cat's eyes are covered with the membrana nictitans while it's being petted, during social grooming, eating, defecating, and so on.

This kitten's expression, and its twisted-back ears, show that it is anxious.

Q.   Why do cats purr?

A.   This is a good question that's difficult to answer. We're not exactly sure why cats purr. We know that kittens first purr while they are nursing. They stop swallowing for a second and purr as they breathe in and out. Their purr may be a signal for their mother to nurse them and let her milk flow. Purring later becomes a signal for a friendly greeting to encourage contact from you or another animal.

There are three categories of sound made by cats:

1. the friendly state of purring we just described;

2. sounds such as "meow" which ask for food, attention, care, or act as a signal that the cat is frustrated; and

3. intense sounds used when a cat is attacking, defending itself or mating. These sounds include hissing, growling, screaming and so on.

What is this kitten yowling about?

Q. *Our five-month-old kitten who plays outside is pregnant. How could this happen to such a young cat?*

A. Unfortunately, female cats become sexually mature anytime from three and a half months to nine months. In males puberty occurs at seven months. Females born in February can have their first heat in May or June, and this is probably your kitten's case. Females born in April usually don't come into heat until the following year.

A cat's heat, or estrus, lasts for three weeks. During that time there are only four to six days in which she can actually get pregnant. However, this is a good enough reason to keep her inside during this time. Some cats, if they do not breed with a male, may have repeated heat cycles and cry or call constantly to get outdoors. Spaying will eliminate this problem.

Q. *How do cats mate?*

A. The courtship behavior of males is triggered by the odor given out by a female cat in heat and by her rolling, courtship display and loud calls. Tomcats may fight. They then display themselves for courtship to the females. This may trigger the female sex organs to mature the eggs and prepare them for fertilization. The eggs are actually released when the male cat enters the female and stimulates her vagina. Before sexual intercourse begins, the male will investigate the female's vaginal area. The female then lowers her front and lifts her hindquarters for the male to mount and penetrate.

The couple may mate several times in an hour. The male cat may become exhausted but if a new female is introduced, he will recover and mate again. Some females choose a particular male to mate with while others have no preferences.

This Siamese is rolling in a courtship display.

Q.   What will sterilization do to my cat? I've heard it makes cats lazy and fat.

A.   This is not true. Castration for males and spaying for females have the same effect on both cats and dogs. When the operation is done at the proper time (just before the first heat in the female and after sexual maturity in the male), it allows the animals' sexual organs to grow to full maturity. Their behavior and weight remain unchanged. "Fixing" your cat cuts down on the cat's desire to roam and fight, especially in males.

Spaying a female cat is a routine operation. The veterinarian removes the reproductive organs—the uterus and the ovaries. In male cats the testicles are removed. Both these operations are done under anesthesia and are painless.

It is a real kindness to do this operation, since it prevents the birth of unwanted kittens. Unsterilized cats also spray all over your home. They caterwaul or scream to be mated when the female has her heat, or when a male smells the scent of a female in heat and wants to go out.

When castration or spaying is done late in a dog's or cat's life, the pet may still want to roam and may act sexually. The hormone-producing organs that are removed by sterilization are of course no longer working, but the animal's behavior pattern has already been established.

Q. Our cat is about to give birth. How can we help her?

A. The best advice, generally, is to let her alone. You can give her a box lined with newspapers and place it in a quiet, dark, warm area of your house, such as a closet. If she won't stay there, let her go where she feels most comfortable. Don't interfere with the delivery. She had been pregnant for between sixty-three and sixty-five days and this is her moment. Watch her delivering the fetuses if you can, though, because it is fascinating! You'll be witnessing one of nature's most beautiful events. However, never let a cat have kittens just so you can enjoy the whole process. It is wiser today not to bring more kittens into this overpopulated world.

If your cat is very attached to you, she may need you nearby to reassure her. First, you'll see a lot of liquid emerge from your cat's vaginal area. This is the bag of water breaking. (Water has surrounded the fetuses during the pregnancy and protected them.) Next the first kitten emerges. The mother cat licks up the water and then she begins licking the kitten. This stimulates the kitten's breathing. After each kitten is born, the mother eats the placenta, the organ through which the fetuses have been fed during pregnancy. She will also cut the umbilical cord, a remarkable example of instinctive behavior. Within an hour of delivery the kittens should be nursing happily. If a newborn kitten gets up and starts wandering away, the mother touches it with her paw or licks it to show it the way back to her nipples.

Q. *It seems that each kitty in our new litter goes to the same nipple each time it nurses. Is this true of all litters?*

A. Kittens do seem to have a nipple or teat preference and to stay with that nipple for weeks. This reduces competition among the kittens and provides good stimulation for continued lactation. The mother cat keeps as clean a home as ever and licks up and eats the kittens' droppings. This will not make her sick, and it is a common practice among mammals.

There are three phases of nursing. In the first, which lasts about two weeks, the mother encourages the kittens to nurse. In the second, the mother and kittens go to each other for nursing. In the third, at the end of three to five weeks, the kittens beg to be nursed. In the wild, by the time the kittens are five weeks old their mother is bringing them rabbits or mice she has just killed. Or she may take the kittens out to hunt. Your cat may also try this in order to stop her kittens from nursing. Weaning may take several months.

Q. *My mother thought our cat Candy was having a false pregnancy after her first heat. Candy grew very large and then after about a month went back to her normal size. What was this behavior all about? What was Candy saying?*

A. Candy was probably saying "I have a desire and instinct to be pregnant, and I want to have a litter of kittens." She very likely did have a false pregnancy. It would be best to spay her to stop her frustration. Some pseudo-pregnant cats adopt a toy or other object and carry it around, licking it and sheltering it like a kitten.

There is also the possibility that Candy really was pregnant, and the fetuses could have been reabsorbed into the uterus.

Q. *We found our cat when she was a three-week-old kitten. Here is our problem. She will not kill the rats in our barn. She chases and catches them but won't kill them. How can we teach her?*

A. Your kitten probably never had the chance to learn from her mother how to bite and kill prey. The cat was probably separated from the litter when she was too young. There may be a critical period early in life for kittens to learn prey killing. However, some cats are under more "instinct control" than others. These cats never need to learn how to kill prey by observing others, knowing what to do instinctively.

When a cat kills, it bites into the neck of its prey and severs the spinal cord. This is probably an innate or inherited response. The series of body movements before the killing bite goes as follows: hunting and searching; running; crouching; stalking; pouncing; grabbing with the forepaws; pinning; and then the killing bite.

When your cat grows older and more confident, she may start killing rats, mice, birds, rabbits, moles, lizards and so on. Some cats can even catch fish from a pool or stream. If your cat isn't a working farm or warehouse cat, whose job it is to keep pests away, it might be kinder to keep her indoors. Wildlife needs to be protected from house cats who kill for fun, not for survival, because such cats already have homes and good food.

This cat's mother never taught it that mice were enemies.

Q. What is my Siamese cat Jewel saying to me when she brings me birds and mice? She's just had her second heat and has not had a litter.

A. These dead prey are not always gifts, as many people think. Since Jewel had no kittens, she may be saying she wants some. She may be acting out a motherly behavior of bringing dead prey for her young and showing them that this dead prey is food. Spaying might help Jewel. However, neutered cats of either sex will sometimes bring prey home, and this may well be a token "gift" for their owners.

Q. My yellow cat loves to sit on a sunny windowsill. When she is there, Sunflower wags her tail and gets excited. She even chirps a little.

A. Sunflower is probably seeing a bird or some other living thing outside. Tail wagging is body language displaying excitement, and together with the chirping sound could mean "I see some prey"—a mouse, bird or insect.

Q. Why do kittens and cats play so much?

A. Play serves many functions for a cat's body language and behavior patterns. When kittens or cats play—jumping up on counters, hiding behind couches—they come in contact with their surroundings. They learn more about their environment. They get to know each other, establish relationships. When a cat plays outside, it finds spots where its prey hide, in little holes, in barns and in trees. Encourage your cat to play with you. It will be more intelligent and more responsive to you. Kittens and cats love to play inside a cardboard box or paper bag, and a super prey-play toy is a ball of paper or wool attached to a length of string that you pull so that your cat can pounce, chase and bite. Some cats even enjoy retrieving.

This acrobatic kitten is amusing himself by hanging from a curtain.

All kittens are curious.

There are many varieties of cats. The all-American short hair tiger-stripe is one of many that makes a fine pet.

Which kitten is best for you? There are tests you can do to help you decide.

5.
Bringing Up Your Kitten

If you don't already own a cat but are thinking of getting one, there are a few facts about breeds and kitten body language that you should know.

First, the breed. If you want a thoroughbred, make an effort to read about the different breeds of cats. Certain characteristics run true in different breeds. For instance, Siamese cats tend to be more dog-like, meaning they are more trainable and responsive to their owners. Long-haired cats such as the Angora or the Persian naturally need more care since their coats must be brushed quite often.

Second, observe the mother's body language carefully. Does she run away and hide from you, saying, "I'm fearful, don't come near me," with her movements? Does she greet you and allow you to pet her? Notice what she's saying because her temperament can be passed on or inherited by her kittens. It's best to avoid kittens from the litters of a fearful mother.

Third, watch the body language of the kittens as they play and interact with each other. Avoid the very dominant or very timid kittens. Both extremes are poor traits for a household pet. Consult the following questions for some ways to test a kitten for the presence of good traits.

When you bring your chosen kitten home, show it around its new household. Introduce it to all the members of your family and show it its litter tray and where it will sleep. Pet and reassure your new kitty often. This will make your bond with your pet a strong and lasting one. Don't protect your kitten too much, however. A little stress and discipline in the life of a cat will give it an environment in which it can learn more easily. Overprotecting smothers independence and curiosity.

Last, provide your kitten with toys, a scratching post and a litter box, good nutrition and lots of attention and affection. If you go away for a while, have someone come in to feed your pet and give it water every day, and to change the litter as needed. Cats would usually rather stay home than travel to new places. If you are at school all day and no one is at home, think about getting another kitten. Two are more fun and will keep each other company. Enjoy your pet while it is a kitten, for it is especially lovable while it is young. Watch its body language as it grows older. More fun is in store.

Questions

Q.   What are the tests I can use to tell which kitten would be a good pet? I want to get my kitten soon.

A.   The first test you can try is picking up a kitten. If a kitten is only three or four weeks old, it is not mature and cannot be judged entirely accurately. A kitten should be about six weeks old when you take it home. Watch out for much older kittens, however, for if they have not had a great deal of contact with people at an early age, they may make poor pets.

When the kitten is over six weeks, you can get a good idea of how friendly and adaptable it is. When you pick it up, it may claw or bite at first. It should, however, soon relax in your arms. Don't take the kitten if it remains fearful and rigid or continues to claw or bite.

The second test is to take your favorite kitten into a darkened room and let it explore. If it freezes and does not eventually begin to sniff around and examine its new surroundings, it may be too timid to make a good pet. Call the kitten and try to get it to follow you. If it ignores you, the cat may not be "socialized" or sufficiently emotionally attached to people to enable it to be a good pet.

The third test should be done in an area the kitten is familiar with. Place the kitten on the floor and throw a ball of paper or knotted string for the kitten to catch. Does it run after the object when you throw it? If the kitten is not willing to play, it may not make a good pet.

Try all three of these tests before you make your final choice. They provide very good indications about whether or not you have picked out a responsive, playful and friendly pet.

Q.   How will our older tomcat Tuffy react when we get our new kitten? How should we introduce them?

A.   When you get your new kitten, give Tuffy lots of attention and treats. He was there first. Make him feel very secure. The same would go for a dog too.

Children often become upset when a new baby arrives. This might happen with your tomcat too. Don't overprotect the new kitten. It will learn a lot from Tuffy.

Q.   Are toys good for a cat?

A.   Yes. They're great for a cat—and for dogs too. They encourage a cat to play and help it to learn.

Q. How should I handle our kitten?

A. Touching and handling your kitten is very important for its emotional growth. When an owner doesn't touch a kitten, especially during the important period of four to eight weeks of age, it may become wild and impossible. As we said, if its mother refuses to touch it also, it may even die.

Pet and cuddle your kitten a lot. Not too much, of course. The kitten will lose its fear of strangers in the future. Also groom your kitten every day or so with a soft brush. This will help it keep a clean and healthy coat.

Q. My kitten bites so hard. I know you can't discipline cats, so what should I do?

A. You can and should discipline cats. Cats are often unruly and destructive only because people think they can't be trained. When you expect a cat to be untrainable, or aloof and distant, the cat often turns out that way.

To discipline your kitten, imitate a mother cat's body language. If the kitten bites its mother, she cuffs the kitten on the nose. You can do the same—but very gently, of course, so as not to hurt it. Tap the kitten gently with one finger, one or two seconds *after* saying "NO" very firmly. That way it will soon learn what "NO" means. Begin discipline as soon as you get your kitten so it will learn right away. Cats who are brought into your home must not rule it and destroy things. The cat has to learn to relate emotionally to others and to become socialized.

If your kitten is long-haired, teach it to stand still for grooming while it is still very young. Groom it often to make the cat familiar with this activity.

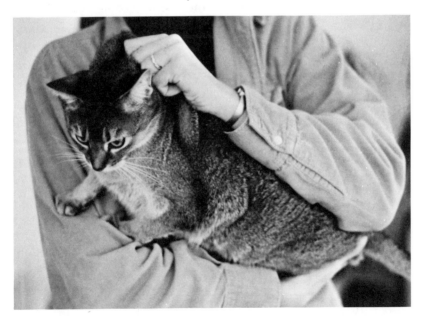

You can hold a cat by the scruff, or loose skin on its neck, just as a mother cat holds her kittens, but you must support its body with your other hand.

Q.   How can I housetrain my male Burmese kitten Sasha?

A.   If only we could housetrain dogs as easily as cats! Simply show Sasha where the kitty box is. A good place for it is the bathroom or a little-used closet or room. When your kitten makes a mistake, tell him that he has done something wrong and show him the box again. Kittens work from instinct and their instincts are to keep immaculately clean and to eliminate in one special place. Clean the box out every few days; cats don't like a very dirty litter box. A sprinkling of baking soda on the bottom will help cut down the odor.

Q.   How can I make sure my kitten is healthy, and how can I keep her healthy?

A.   Dr. Fox recommends taking your kitten to a veterinarian when you first get it. It is very sad when a young kitten dies from some disease or physical problem. The vet will look for tapeworms, fleas, ear mites and roundworms. The doctor will also give your kitten a series of shots to protect it from three very common cat diseases: feline distemper, feline influenza and feline pneumonia. Avoid buying flea collars and powders, coat conditioners and other remedies from pet stores. Only use what the vet prescribes and *never* try to treat your cat with any "quack" remedy when it is sick.

Here's another tip to protect your little kitten when you get it home. Learn to slide your feet along the floor when you walk, and open and close doors slowly and carefully. There have been many tragic cases of kittens getting injured or even killed by a heavy footstep or a slamming door. Also, don't let your kitten roam outside if you live in a populated area. The biggest danger is being run over by a car. Cats can also pick up parasites and diseases from other cats outside and they can get injured in fights. And we mustn't let cats kill beautiful wild birds. Cats can be very happy with a life in your house or apartment.

Q.   *What can I do about my cat? He claws and ruins our furniture.*

A.   You're going to have to get tough and speak his language. Give him a little cuff on the nose, yell "NO" and pick him up and take him to his scratching post each time he scratches the furniture. Help him scratch at the post. A little catnip rubbed into the carpet or scratching surface of the post may make it more attractive. If he continues to ruin the furniture, have a cat groomer show you how to clip his claws carefully. If he *still* persists, you may have to have him declawed, but many cat lovers regard this as an inhumane mutilation.

Although some veterinarians disapprove of this declawing operation, it is approved by many others. The operation involves anesthesia. The claws on the front feet only should be removed. After the operation, the cat should be kept indoors because it won't be able to defend itself so well without its front claws or be able to climb trees.

Q.   *Felix eats a lot of grass. Is there something missing in his diet?*

A.   Most likely Felix is fine, and is just doing what comes naturally. Meat-eating animals in the wild often eat the stomach of their prey first. The stomach contains all sorts of plants and grass material. Grass is fine for cats and dogs. It may also help the cat vomit up any fur balls it has swallowed.

Q. *What is a good diet for a cat? Mine likes one commercial cat chow and table scraps.*

A. It sounds as though your cat has a well-balanced diet, providing the food is not all protein. Dr. Fox feeds his cats—a Burmese and an Abyssinian—dry cat food, which they are free to snack on all day. Morning and evening he gives them moist canned cat food and any table scraps (one cat is particularly fond of corn on the cob and peas). Give your cat a well-balanced diet high in good-quality protein, plus a few vegetables, fruit and cereal, if it enjoys it.

Be careful of the "all-meat" diet some commercial cat food companies push, unless it says on the can or package that it is a complete and balanced diet. These could kill your cat. Your cat may need a special diet if it is old or very young or has a kidney infection (to which cats are particularly prone) or is pregnant. Ask your veterinarian for advice as well.

Q.   *My Persian cat Nima developed a rash on her neck, chest and back. We gave her the drug cortisone, as the vet prescribed, and it went away but now it has come back. What is the problem?*

A.   Nima might have an allergy to food, fleas or some furniture polish. Ask your vet again. Too much cortisone is dangerous and the problem has to be resolved.

Q.   *My nine-year-old cat has blood in his stool. What do you think this is from?*

A.   Get your cat to a veterinarian as soon as you can. This could be blood from the intestines. Or there could be some problem around the anus, or even cancer. However, constipation is one of the most common causes of this problem. A little raw milk added to moist canned food may help keep the stools loose.

Q.   *Cindee, our beautiful feline, has scratched herself so hard around the ears that she's bald there. What do you think she is saying to us?*

A.   Her behavior is a sign that she has a serious ear problem. She probably has ear mites. These microscopic parasites are picked up outside from other cats or from the cat's mother while it is still a kitten. They cause continual irritation, and lots of dark brown ear wax is produced. See your vet to get some medication for this problem.

6.
Fun and Problems

What is your cat saying to you when it suddenly becomes unhousebroken and leaves its droppings everywhere but in its kitty litter? It's saying, "Something's wrong with the way we're living." Perhaps you have moved recently (cats are often even more attached to their living place than their owners). Or perhaps there's a new pet in your house and the cat is upset with all the attention the newcomer is getting.

Cats have fewer behavior disorders than dogs. This is because cats are naturally more independent while dogs are dependent upon a leader, usually their owners. Dependency on others in dogs and humans can cause problems. Independence in both animals and humans is desirable because it shows emotional maturity, an ability to survive on one's own. But not all cats are free from problems. In this last chapter we'll explore what problems and behavior disorders cats do have.

Your cat's problems may arise from the temperament with which it was born, or they may be due to the way it has been raised. No cat will be a loving, affectionate pet if it has had no affection and care from its owners. And if it is allowed to claw furniture and eat house plants in its own house, there's no reason for it to stop this destructive behavior when it goes to stay with Aunt Betsy.

Another reason cats may develop behavior disorders is if there is a change in their relationships—if its owner has died or a new baby has arrived in the household. The cat may start refusing its food and become depressed.

One cat reacted violently to its owner's death and wouldn't let anyone near the owner's body at first. After the funeral the cat would go every day to visit his masters' grave.

These reactions are extreme but they are behaviors you should know about to understand what your cat is saying all through its life.

Questions

Q. My ten-year-old cat always grooms and licks himself before a thunderstorm. He's been doing this for as long as I can remember. Will you please give me an explanation for this weird behavior?

A. This grooming behavior before a thunderstorm is a sign that your cat is nervous. Your cat can sense the oncoming storm and he is upset. The urgent licking activity is taking the place of his nervousness. It is called displacement, and many mammals, including humans, make such nervous gestures. Do you ever bite your nails when you're nervous or flick your hair out of your eyes? Birds peck and preen when they're ambivalent about approaching another bird. Dogs frequently scratch themselves, and cats will suddenly sniff the ground when they are embarrassed or "uptight," especially when facing a rival.

Q. Why does my cat groom himself until he's caused a bald spot?

A. Your cat is probably bored. Boredom makes cats overgroom themselves. This often happens in zoos where animals are bored or frustrated. Carefully check the animals, especially the lions and the tigers, the next time you're in a zoo. In many zoos these poor captive animals don't have any room to run or do much of anything, so self-mutilation can result from constantly licking or sucking on a paw or a tail.

Your cat's body language may be saying "Why don't you play with me more, or at least buy me some toys or another pet?" Try to play with it or provide it with toys. Disciplining this behavior is important too.

Q. My Siamese cat Karim kneads, chews and sucks on my good wool sweaters. Why on earth does she do this? She's ruined three already.

A. Karim has to be disciplined. Tell her "NO" and cuff her on the nose. Siamese cats have a special attraction to wool. Perhaps the lanolin has an odor like a mother cat's nipple and kneading, chewing and sucking is the cat's way of nursing. Some cats also sniff under their owners' armpits and try to nurse.

Q. My cat Coco had a fight with a dog. When he came home from the fight, Coco was panting, salivating and sweating from his footpads. He hid under the couch and refused to come out. His pupils were dilated and he was shy and shivering. Coco was in such severe shock I called the vet. He said to leave him alone and he should recover. Why was Coco so terrified?

A. Cats have extremely sensitive nervous systems and they often overreact to unexpected stimuli such as this fight with a dog. Sometimes a phobia will result and your cat will continue hiding. Handle your cat gently in this situation. In severe cases a vet would have to give a tranquilizer. Once in a while, if the shock is serious enough and the cat is very high-strung, it will die. Siamese cats are very vulnerable to shock, sometimes even when they're being held down by a veterinarian. A tranquilizer will help calm these nervous cats.

Q. *When my cat Quincy saw a huge Doberman outside on a leash held by its owner, she froze and didn't want to move. Why?*

A. Poor Quincy got quite a shock seeing this huge enemy. She was so alarmed she became catatonic, unable to move at all. We don't know just why animals do this but it may serve to keep predators from chasing them. Sometimes predators won't run after a still prey. That's why you see rabbits or mice "freeze" at times. Their running triggers their predators' chase instincts, as we mentioned in Chapter 4. *Not* running away could help save a cat's life.

Q. *I think my Abyssinian cat has a behavior disorder. He comes to sit on my lap, rolls over and purrs, but if I try to tickle his stomach, he meows, scratches and runs away. Isn't this strange?*

A. Not really. Tickling a cat while it's on its back seems to release a defensive-aggressive reaction. This is a posture in which a cat strikes out against an aggressor. So for the moment its body language is saying "I'm your enemy."

Q.   Why do our kittens act so wild at night? They chase and fight each other regularly for at least an hour every night.

A.   Cats are nocturnal animals and "switch on" at night. Many have what Dr. Fox calls "the evening crazies." Join in with them; they may well enjoy playing Chase and Hide-and-Go-Seek with you.

Q.   My cat attacked a friend who came to visit. Why? What happened? She's always been friendly with us.

A.   Cats often have a dog-like reaction to the invasion of their territory. A cat will often spray a visitor's overnight bag if it feels its territory has been invaded or violated. The cat may refuse its food, leave home or even attack as yours did. She may also be jealous. Give your cat extra attention when someone comes to visit. Indulge her with treats until she gets used to the visitor.

Q.   Why does my pet cat Button have to be put to sleep? He's fourteen and has been sick for a while. The veterinarian told us he's in a lot of pain, and he does cry when he gets up sometimes. It also seems to hurt him when he walks. The vet told my Mom it was time to put Button away, but I can't let them do this. Can't we wait and see if he'll get better?

A.   It is probably time to put Button to sleep. Your vet has seen that Button's condition is causing him too much pain and suffering. *You* also noticed his body language. It is saying "I'm very sick and hurt a lot. Please help me." The doctor has probably done all he or she could do to help Button get better and now knows there is little hope for his recovery.

Think of all the happiness you and Button have given each other. You have given him a nice home, fed and cared for him. You love him very much and now you can show even more love by humanely ending his suffering. Your veterinarian can painlessly inject an anesthetic into the cat's vein so that he will quietly go to sleep.

Button will always have a special place in your memory. Cherish this memory and remember your love for him. Your care and love for your cat helped you become a better person as much as it gave Button a happy life.

Q. *I had a dog who died a year ago. I'd like to get a cat but you can't really train a cat to do tricks, can you?*

A. Dr. Fox has trained his Abyssinian and Burmese cats to "come," to chase toys and bring them to him and to follow him on walks. And you don't have to be a doctor to do this. Simply train your cat the way you would a dog. Whenever your cat obeys your commands, give it treats and warm petting. You may have to work harder than with a dog. Some cats (and some dogs) refuse to be taught and there's not much you can do about reaching them. Siamese cats are more trainable than other breeds. They are the most dog-like of the cat breeds. Cats can be easily leash-trained when they are young and enjoy going out for walks.

You can also try out your cat body language on your new kitten. Rough-and-tumble with it. Cuff it playfully and gently and allow it to cuff you back. Play Hide-and-Go-Seek.

Q. *I know people can be like dogs—we stay in groups and follow our leaders. But is our behavior at all similar to cat behavior?*

A. Many people, like cats, are highly individualistic and want to do things their own way, away from groups. But at the same time we are social animals, like dogs, with social needs. We need acceptance from others—our parents and friends. We might need recognition, status, success and power. So we see that in order to fill these needs, we must give up some of our own independence. As solitary and independent as cats are, they still need each other and us.

Q. *I love cats and most other animals. Where can I learn more about helping them and about careers working with animals?*

A. Probably the best idea is for you to join the Kindness Club, the junior division of the Humane Society of the United States. You will receive a regular magazine and other material that will inform you about all kinds of animals and will help you to understand and help them in many ways. For more information, write to:

>"Kind" Magazine
>2100 L Street N.W.
>Washington, DC 20037

Index

aggressive behavior, 35, 38, 40, 74
Angora cats, 59

balance, sense of, 21, 22
behavior disorders, 70, 72–76
behavior patterns, 35–55
birth, 50
biting, 65
blood in stool, 69
bobcats, 17, 32
body language, 7–8, 9, 10, 11, 21, 35–55
  of kittens, 59
boredom, 72–73
brains, 21, 22

calico cats, 20
castration, 37, 41, 49–50
cats
  behavior disorders of, 70, 72–76
  behavior patterns of, 35–55
  compared with dogs, 7–8, 9, 10, 27
  compared with humans, 78
  ecology of, 17, 18
  emotions of, 21
  evolution of, 11, 17–19, 21
  eyes of, 23–24
  health problems of, 66, 68–69
  history of, 11–16
  homing ability of, 21, 31
  senses of, 21–34
  superstitions about, 15, 16
  in the wild, 8, 10, 38, 51
cheetahs, 18
civets, 19
clawing, 41, 67. *See also* scratching
cleanliness, 16, 66
clouded leopards, 19
courtship behavior, 48. *See also* mating behavior

deafness, 20
declawing, 67
defensive posture, 42, 45, 74

diet, 68
direction, sense of, 21, 30, 31
discipline, 60, 65, 73
diseases, 66
distemper, feline, 66
dogs
  compared with cats, 7–8, 9, 10, 27
  mistreatment of, 9
  in the wild, 10
domestication, 8, 13
dominance behavior, 35, 37–38, 40, 45

ear mites, 66, 69
ears, 20, 27
ecology, 17, 18
Egypt, cats in, 13
emotions, 21
estrus (heat), 48
evolution, 11, 17–19, 21
extrasensory perception, 29
eyes, 21, 23–24

facial expressions, 46
fearful behavior, 35. *See also* defensive posture
fear of cats, 32
fighting, 37, 73
flea collars, 66
Flehmen response, 27, 46
Fox, Dr. Michael, 8, 29
"Freezing" behavior, 74
fur balls, 34, 67

grass, eating of, 67
greeting behavior, 44
grooming, 65, 72–73

health and health problems, 66, 68–69
hearing, sense of, 21, 27–28, 29
heat (estrus), 48
hissing, 7, 47
history, 11–16
homing ability, 21, 31
housetraining, 66, 70
Humane Society, 78

hunting and prey, 13, 17–18, 24, 27, 28, 35, 52–54

inbreeding, 20
influenza, 66
innate behavior, 10, 52
inside vs. outside, 66

Kaffir cat, 11
Kindness Club, 78
kittens, 59–69
   choosing, as pets, 59, 61
   health problems of, 66, 68–69
   and older cats, 63

learning, 10, 28, 52, 54, 60
lions, 10, 17–18, 45
litter box, 66. *See also* housetraining
long-haired cats, 59, 65
lynxes, 20

marking behavior, 35, 37, 41
maternal behavior, 35, 50–52. *See also* birth; pregnancy
mating behavior, 48
membrana nictitans, 23, 46
meowing, 47
mistreatment, effects of, 9
mountain lions, 7

noses, 26–27
nursing, 51

offensive posture, 42
old age and death, 76
overprotection, 60, 63

parasites, 66
parties, 38
passive posture, 43
persecution of cats, 15–16
Persian cats, 59
petting, 25–26, 59–60, 64
physiology, 21–34
play behavior, 35, 54, 61, 75
pneumonia, feline, 66
pregnancy, 48
pregnancy, false, 52
prey. *See* hunting and prey
psi-trailing, 31
puberty, 48
purring, 47

rashes, 69
relative dominance, 35, 37
righting reflex, 22
rolling over, 43, 48
rubbing, 41, 45

saber-toothed tiger, 19
scent glands, 41, 45. *See also* marking behavior
scratching, 69. *See also* clawing
scratching post, 60, 67
senses, 21–34
sexual behavior, 35, 38, 48–50
Siamese cats, 20, 73
sight, sense of, 21, 23–24
smell, sense of, 21, 26–27
sniffing, 45, 73
social investigation behavior, 35, 38, 45–46
solitariness, 18, 21, 35
sounds. *See* vocalizations
spaying, 48, 49–50, 52, 54
spraying, 30, 41, 76
staring, 42, 46
sterilization, 49–50. *See also* castration; spaying
submissive behavior, 43, 45, 46
superstitions, 15, 16
swimming ability, 32

tail postures, 44, 54
taste, sense of, 21, 34
temperature, sense of, 21
territoriality, 30, 38, 41. *See also* marking behavior
tests, for kittens, 61
time, sense of, 21, 29, 30, 31
touch, sense of, 21
toys, 54, 60, 63
training, 77. *See also* discipline; housetraining
trapping, 18
treading, 45

vocalizations, 35, 47
vomeronasal organ, 27

whiskers, 33
wildlife, protection of, from cats, 52–53, 66
witchcraft, 15